Cambridge Elements ☰

Elements of Paleontology
edited by
Colin D. Sumrall
University of Tennessee

FUNCTIONAL MICROMORPHOLOGY OF THE ECHINODERM SKELETON

Przemysław Gorzelak
Polish Academy of Sciences

Paleontological
S O C I E T Y

CAMBRIDGE
UNIVERSITY PRESS

CAMBRIDGE
UNIVERSITY PRESS

University Printing House, Cambridge CB2 8BS, United Kingdom

One Liberty Plaza, 20th Floor, New York, NY 10006, USA

477 Williamstown Road, Port Melbourne, VIC 3207, Australia

314–321, 3rd Floor, Plot 3, Splendor Forum, Jasola District Centre,
New Delhi – 110025, India

79 Anson Road, #06–04/06, Singapore 079906

Cambridge University Press is part of the University of Cambridge.

It furthers the University's mission by disseminating knowledge in the pursuit of
education, learning, and research at the highest international levels of excellence.

www.cambridge.org
Information on this title: www.cambridge.org/9781108810319
DOI: 10.1017/9781108893886

First published 2021

A catalogue record for this publication is available from the British Library.

ISBN 978-1-108-81031-9 Paperback
ISSN 2517-780X (online)
ISSN 2517-7796 (print)

Functional Micromorphology of the Echinoderm Skeleton

Elements of Paleontology

DOI: 10.1017/9781108893886
First published online: January 2021

Przemysław Gorzelak
Institute of Paleobiology, Polish Academy of Sciences

Author for correspondence: Przemysław Gorzelak, pgorzelak@twarda.pan.pl

Abstract: Echinoderms elaborate a calcite skeleton composed of numerous plates with a distinct microstructure (stereom) that can be modeled into different shapes thanks to the use of a transient amorphous calcium carbonate (ACC) precursor phase and the incorporation of an intraorganic matrix during biomineralization. A variety of different types of stereom microarchitecture have been distinguished, each of them optimized for a specific function. For instance, a regular, galleried stereom typically houses collagenous ligaments, whereas an irregular, fine labyrinthic stereom commonly bears muscles. Epithelial tissues, in turn, are usually associated with coarse and dense stereom microfabrics. Stereom can be preserved in fossil echinoderms and a wide array of investigating methods are available. As many case studies have shown, a great deal of important paleobiological and paleoecological information can be encoded by studying the stereom microstructure of extinct echinoderms.

Keywords: Echinodermata, biomineralization, stereom, function, paleoanatomy

ISBNs: 9781108810319 (PB), 9781108893886 (OC)
ISSNs: 2517-780X (online), 2517-7796 (print)

Contents

1 Introduction

Although data on echinoderm biomineralization mostly come from studies on echinoids, it is widely accepted that all echinoderm classes build their skeletons using the same method of biomineralization, as inferred from very similar structures, properties, morphogenesis and organic molecules involved in the skeletogenesis (Dubois and Chen, 1989). Echinoderm skeletons are formed within the syncytium (i.e., a multinucleate cell) through a biologically controlled intracellular biomineralization process (e.g., Okazaki, 1960; Märkel 1986; Weiner and Addadi, 2011). They are composed of many calcite plates with a unique trabecular microstructure called stereom: herein lies the major synapomorphy of Echinodermata (e.g., Bottjer et al., 2006). In this Element, I summarize current and state-of-the-art knowledge of echinoderm biomineralization, with particular attention to the stereom microstructure and its functions. I also discuss methods for investigating stereom in fossil echinoderms, and highlight through case studies the great potential of micromorphological observations in determining the paleoanatomy and paleoecology of extinct groups of echinoderms.

2 Biomineralization, Structure and Biogeochemistry of the Echinoderm Skeleton

Larval and postmetamorphic skeletons of echinoderms are of both internal and mesodermal origin. However, they can be exposed to the external environment if the syncytial membranes are degraded (e.g., Märkel, 1986). The larval skeleton is composed of spicules made of a few rods that are almost completely resorbed at the end of the larval stage. The postmetamorphic skeleton, in turn, consists of numerous ossicles bound together by connective and/or muscle tissues that begin to form *de novo* just before settlement and metamorphosis (Hyman, 1955).

Sclerocytes (the so-called skeleton-forming cells, SFCs) along with odontoblasts (which are involved in the tooth formation) have long been considered the only cell types directly involved in biomineralization (Ameye, 1999). However, it has been recently shown that even nonspecialized epithelial cells may also be involved in this process (Vidavsky et al., 2014). Additionally, the role of phagocytes and/or spherulocytes in biomineralization has been recently hypothesized (Kołbuk et al., 2019). The formation of the skeleton takes place within vacuoles or vesicles that are typically enclosed in a syncytial pseudopodium formed by SFCs. Initially, the skeleton is being formed intracellularly and then extracellularly. Several ion transport systems, including special ion pumps, channels, exchargers and cotransporters, have been suggested to play a key role in echinoderm biomineralization (for a review, see Dubois and Chen, 1989). It has been also suggested that organic components may be involved in this transport (Beniash et al., 1999; Wilt, 1999). Furthermore, as recently shown by Vidavsky et al. (2016), seawater with its ions can be directly incorporated into the cells of echinoid embryos by endocytosis. However, the exact pathway of ion transport into the calcification site in adult echinoderms is still uncertain. A generalized crystallization pathway in echinoderms has been proposed by Weiner and Addadi (2011; see also Figure 1A). According to this model, once ions have been transported to the vesicles, formation of the first disordered mineral phase (amorphous calcium carbonate ACC) initiates. Subsequently, the vesicles bearing ACC nanoparticles with occluded biomolecules are transported into the syncytium where translocation and secondary crystallization of nano-grains (from disordered into a more ordered phase) proceeds on the organic template (for more details, see Weiner and Addadi, 2011; Seto et al., 2012).

The mature skeleton of echinoderms is composed of magnesian calcite comprising small amounts (up to about ~10 wt%) of stable amorphous calcium carbonate (ACC), water (< 0.2 wt%), and up to about ~1.6 wt% of intrastereomic organic matrix (Alberic et al., 2019). $MgCO_3$ content may vary widely from about 3 to 43.5 mole% (e.g., Schroeder et al., 1969). Skeletal Mg/Ca ratio in echinoderms can be affected by a number of environmental and physiological factors (such as ambient seawater Mg^{2+}/Ca^{2+} ratio, temperature, salinity, diet and "vital effects"; for a review, see Kołbuk et al., 2019; Kołbuk et al., 2020).

Under X-ray diffraction and polarizing microscopy, any individual ossicle behaves as a single calcite crystal (e.g., Donnay and Pawson, 1969; Yang et al., 2011). However, the crystal texture (i.e., coherence length [the average distance between imperfections in specific crystallographic directions] = 50–250 nm, and angular spread of perfect domains [degree of misalignment between perfectly coherent domains] = 0.1°) differs significantly from those observed in abiotic calcite (usually 500–800 nm and 0.01°, respectively). Furthermore, the crystal texture may differ among the different fabrics of the stereom in the same ossicle (Berman et al., 1993; Aizenberg et al., 1997). For instance, it has been shown that the perfect domains are more isotropic in the trabeculae of the open, regular stereom than in the lateral septa composed of imperforate stereom layers where coherence length is lower in the direction perpendicular to the *c*-axis (Aizenberg et al., 1997). Additionally, when broken, the echinoderm skeleton displays conchoidal fracture that reduces the brittleness of the material by dissipating strain energy and deflecting crack propagation (Berman et al., 1988). This unique property of stereom, that is generally harder, stiffer and stronger than inorganic calcite, has been ascribed to the involvement of organic molecules (the so-called intrastereomic organic matrix; IOM) in the

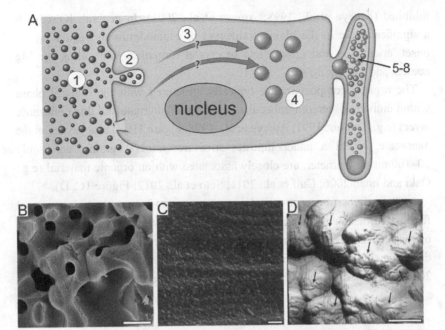

Figure 1 Biomineralization in echinoderms. A. Biomineralization pathway (redrawn and slightly modified after Weiner and Addadi, 2011): (1) the medium (seawater and/or body fluid) from which the ions are derived, (2) the ion-sequestering via endocytosis of seawater droplets and/or ion channels and/or transporters, (3) transport within the cell to vesicles, (4) vesicles in which the first disordered ACC phase forms, (5–8) transport of the ACC-bearing vesicles into the syncytium, translocation of the disordered ACC phase to the crystallization front and its transformation into more ordered phase. B. Fully mature mineralized tissue showing a concentric lamination revealed by slight acid etching (FESEM [Field Emission Scanning Electron Microscopy] micrograph). C. Magnification of the fractured stereom bar revealing nanogranular structure (FESEM micrograph). D. Magnification of the fractured stereom bar revealing nanograins with possible organic envelopes (arrows) (3D visualization of the height mode AFM [Atomic Force Microscopy] image). Scale bars: 10 μm (B), 200 nm (C), 100 nm (D), respectively.

biomineralization and their incorporation into the skeleton at different structural levels (e.g., Weiner, 1985). These organic components include various proteins and glycoproteins (N-glycoproteins, O-glycoproteins and terminal sialic acids), distribution of which may be variable, that is, N-glycoproteins are preferentially located in the putative amorphous subregions of the stereom whereas O-glycoproteins are localized in the subregion where skeletal growth is inhibited (Ameye et al., 1998; Ameye et al., 2001). In general, IOM plays a significant role in the biomineralization of echinoderms by regulating the onset, orientation and growth rate of crystal formation, controlling the Mg content and stabilizing ACC.

The organic components are typically structured with the mineral phase within individual stereom trabeculae in a form of alternate and thin concentric layers (e.g., Dubois, 1991; Ameye et al., 1998; Figure 1B). These layers, at the nanoscale, reveal a nanocomposite structure: mineral grains, commonly 20–100 nm in diameter, are closely associated with an organic material (e.g., Oaki and Imai, 2006; Cuif et al., 2011; Seto et al., 2012; Figure 1C, D).

At the microscale, each ossicle appears in the form of a tridimensional fenestrated meshwork of trabeculae, the so-called stereom. Extensive studies on the skeletal microstructure have documented different stereom microfabrics that have similar morphogenesis (Dubois and Jangoux, 1990; Gorzelak et al., 2011; Gorzelak et al., 2014a; Figure 2). Initially stereom trabecular bars grow in

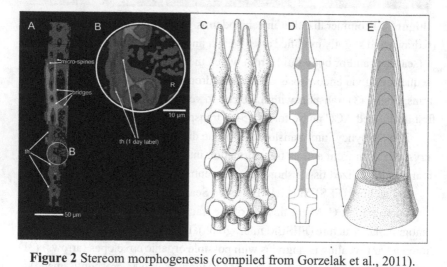

Figure 2 Stereom morphogenesis (compiled from Gorzelak et al., 2011).
A. NanoSIMS maps of the ^{26}Mg/^{44}Ca distribution in labeled spine stereom of *Paracetrotus lividus* Lamarck during 1-day ^{26}Mg labeling event.
B. Enlargements of the stereom showing 1-day thickening process ("th" and arrows) during the ^{26}Mg labeling event on the previously formed stereom bars. R = resin filled pores, blue regions = growth in normal (i.e., unlabeled) artificial seawater with normal ^{26}Mg/^{44}Ca ratio, red-yellow regions = enhanced ^{26}Mg/^{44}Ca ratio due to ^{26}Mg labeling. C–E. Model of stereom formation. Orange regions = distribution of ^{26}Mg label as observed in experiment, thin labeled thickening layer (bottom in D) is continuous with massive labeling.

a form of thin conical micro-spines. The neighboring micro-spines then fuse together by lateral bridges forming a thin meshwork of inner stereom that thickens simultaneously and very slowly by addition of thickening layers (Figure 2). The stereom pores may be eventually filled by secondary calcite to form perforate or imperforate stereom layers (Gorzelak et al., 2017a).

3 Stereom Types and Relationship to Phylogeny, Growth Rate and Investing Soft Tissues

Early work on skeletal microstructure in echinoderms recognized two basic types of stereom: regular (termed alpha) and irregularly arranged meshwork (termed beta) (Roux, 1970). Subsequently, however, Macurda and Meyer (1975) found that the stereom design in modern crinoids may be much more complex. The latter authors stressed that stereom may be quantified and introduced a new terminology for alpha and beta stereoms, which were then referred to as galleried and labyrinthic stereoms, respectively. In his seminal paper on echinoid microstructure, Smith (1980a) identified a variety of different stereom types, and provided their formal definitions (see Sections 3.1–3.8). This author pointed out that the stereom coarseness can be quantified using maximum pore diameter and trabecular thickness (measured in the narrowest point between adjacent pores) (Figure 3). He also stressed that the stereom microstructure can be controlled by three major factors: phylogeny, growth rate and type of investing soft tissue.

Phylogenetic signal in plate microstructure of echinoids was already noted by Nissen (1969) who pointed out that cidaroids in general display much more regular stereom in plates than any other groups of echinoids. Notable differences in the basic pattern of stereom design in plates and spines between different echinoid taxa were recognized by Smith (1980a, 1984, 1990) and Kroh and Smith (2010). Phylogenetic significance of stereom architecture in crinoids has also been emphasized by Simms (2011), who highlighted that the stereom design of columnal latera more likely reflects phylogeny than function. Simms argued that labyrinthic stereom is taxonomically widespread and plesiomorphic, whereas perforate stereom is apomorphic for Isocrinida.

Stereom appearance may be also affected by the rate of calcification. For instance, Pearse and Pearse (1975) found that well-fed and fast-growing echinoids produced their plates dominated by opaque growth zones (composed of regular stereom), whereas starving and slow-growing individuals formed plates composed of translucent zones (having more irregular stereom). Additionally, Smith (1980a) noted that trabecular thickness and porosity may be influenced by growth rate. He argued that rapid growth is commonly associated with open stereom, whereas reduced growth rate is

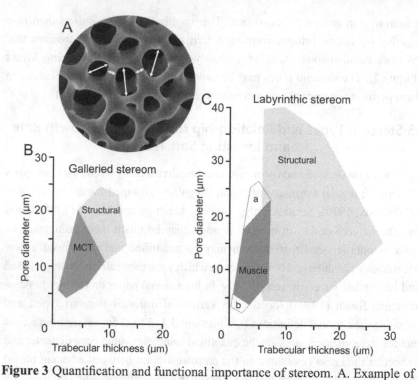

Figure 3 Quantification and functional importance of stereom. A. Example of measurements of pore diameter and trabecular thickness in stereom. B. Galleried stereom of all echinoderms for mutable collagenous tissues (MCT) and structural rectilinear stereom. C. Labyrinthic stereom of all echinoderms showing stereom associated with muscles, area "a" delimits stereom of deep-sea echinothurioid echinoid tubercle associated with collagenous insertion: area "b" delimits stereom associated with either muscle or collagenous fibers (B and C redrawn from Smith, 1990).

correlated with denser stereom microfabrics. Development of new labeling methods using stable isotope ^{26}Mg (Gorzelak et al., 2011; Gorzelak et al., 2014a) or trace element manganese (Gorzelak et al., 2017a) to study growth dynamics and morphogenesis of stereom at the micro- and sub-micrometer spatial resolution, confirmed and specified that the so-called open stereom may indeed grow very rapidly (>100 μm per day), but simultaneously it thickens at a very slow rate (~1 μm per day). Gorzelak et al. (2017a) also demonstrated that denser stereom microfabrics (perforate or labyrinthic constructional stereom) along the inner plate margins in echinoids form much more slowly than open rectilinear or galleried stereom expanding in interradial and meridional directions. The latter authors also showed that the outermost perforate stereom in spines develops much more slowly than the inner rectilinear stereom. Likewise, formation of spine septa composed of imperforate stereom proceeds rather slowly and is a highly complex process involving deposition of porous stereom that is secondarily filled by calcite.

Notwithstanding, the nature of investing soft tissue is the most important factor controlling stereom design. Extensive studies on stereom microstructure of extant echinoderms (Roux, 1970, 1971, 1974, 1975, 1977; Heatfield, 1971; Macurda and Meyer, 1975; Macurda, 1976; Smith, 1980a) identified that each stereom type is commonly associated with a particular type of soft tissue. Smith (1990) pointed out that, by integrating quantitative and qualitative approaches, it is possible to distinguish areas of collagen, muscle fibers and epithelial tissue (Figure 3).

Collagenous fibers binding ossicles together are arranged in deeply penetrating and long straight bundles. As a consequence, stereom associated with this type of tissue typically appears in the form of well-aligned galleries (Grimmer el al., 1984). Muscles in echinoderm ossicles, in turn, usually splay out and are anchored around fine or moderately fine labyrinthic or retiform stereom, which forms a distinct platform growing peripherally. In crinoid arms, this stereom is typically associated with needle-like projections (Macurda and Meyer, 1975). However, if collagenous fibers are not penetrative and short, they may also loop around the trabeculae, superficially resembling labyrinthic stereom. For instance, at autotomy planes in crinoid columnals and brachials, collagenous fibers are short and thin, and anchor around labyrinthic-like stereom laying over galleried stereom (e.g., Gorzelak, 2018). Likewise, small and nonpenetrative collagenous fibers associated with pedicellariae and small spines attach to very fine labyrinthic stereom (trabecular thickness < ~4 µm; pore diameter < ~5 µm) (Smith, 1980a). Another exception is found in some deep-sea echinoids, such as echinothurioid echinoids in which collagenous fibers of their tubercles attach to open (pore diameter > ~17 µm) and fine (trabecular thickness ~2–8 µm) labyrinthic stereom (Figure 3), and in a few stalked crinoids (Hyocrinidae and Holopodidae) in which ligaments may also attach to labyrinthic stereom (Holland et al., 1991; Roux et al., 2002). These labyrinthic-like stereoms associated with short, non-penetrative collagenous fibers, in contrast to thin muscle-bearing labyrinthic or retiform stereom layer, are however usually extended to a thicker layer composed of at least a few levels and/or are more open textured (Smith, 1980a). The exception to this observation is the pillar bridges of echinoid teeth, which are composed of a retiform layer and bear collagenous ligaments. Noteworthy, although muscle fibers in spines of echinoids usually attach to superficial, open and irregularly arranged fine stereom, in certain species of cidaroids, spine muscles may be solely linked to unmodified rectilinear or galleried stereom (Smith, 1980a).

Stereom diagnostic for epithelial tissue is usually dense and thick. It appears in the form of coarse labyrinthic stereom or, most commonly, perforate to imperforate stereom layers. Superficially, these stereom types can be variously ornamented with pegs, domes or thorns. These stereom microfabrics display high stiffness and hardness, and are thought to increase skeletal strengthening and resistance to abrasion (e.g., Moureaux et al., 2010).

In the following, I have compiled data (Roux, 1970; Macurda and Meyer, 1975; Macurda, 1976; Smith, 1980a; 1990; Medeiros-Bergen, 1996; Reich, 2015; Martins and Tavares, 2018; Gorzelak, 2018) and provide a summary of each stereom type, with special references to their respective appearance, distribution and association with soft tissues.

3.1 Rectilinear Stereom

Regular array of trabeculae arranged in cubic (or orthorhombic) lattice. Pores well aligned and equal in three directions, perpendicular (or nearly perpendicular) to one another; their diameter is comparable to the thickness of the trabeculae (Figures 4A–D). It is commonly used for general plate construction (in echinoids and asteroids). It is also associated with penetrative collagenous tissues (in crinoid columnal and brachial articular faces).

3.2 Galleried Stereom

Deeply penetrating parallel galleries running in one direction only. No regular surface pattern. Lateral pores usually smaller than the galleried pores (Figures 4E–H). It is commonly associated with penetrative collagenous fibers (4–12 μm in diameter) running parallel to galleries (e.g., suture faces and boss of large tubercles in echinoids; suture faces in asteroid plates; brachial ligament pits, columnal, and cirral articular faces in crinoids; intervertebral and spine ligament insertion areas in ophiuroids). In some rare cases it may be connected with muscles inserting on outwardly growing surfaces (e.g., perradial muscle pits in asteroid plates).

3.3 Labyrinthic Stereom

Irregular meshwork of trabeculae. Trabeculae and pores form unorganized tangle at any section and may be of different size (Figures 4I–L):

(i) Fine meshwork (pores < ~6 μm, trabeculae < ~3 μm) is associated with attachment areas for fibrous tissues (commonly muscle, rarely collagenous) (e.g., small tubercle bosses and associated spine bases in echinoids, asteroids and ophiuroids; areoles of echinoid tubercles; muscle flanges of ophiuroid vertebrae; tube foot attachment areas in echinoids, asteroids and ophiuroids);

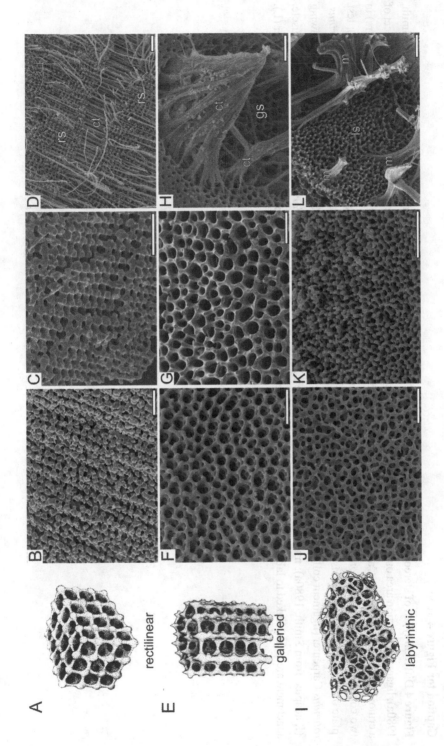

rectilinear

galleried

labyrinthic

Caption for Figure 4 (cont.)

Figure 4 Different types of stereom fabrics in recent echinoderms. A–D. Rectilinear stereom: a schematic drawing (A, adopted from Smith, 1980a); ligament insertion area on a radial of a crinoid *Holopus rangi* Carpenter (B); penetrative collagenous tissue area on columnal facet of a crinoid *Metacrinus rotundus* Carpenter (C); long through-going collagenous ligamentary tissue (ct) penetrating rectilinear stereom (rs) in two columnals of *Metacrinus rotundus* Carpenter (D). E–H. Galleried stereom: a schematic drawing (E, adopted from Smith, 1980a); test plate of a cidaroid *Prionocidaris baculosa* Lamarck (collagenous sutural fiber area) (F); articular facets of a crinoid cirrus of *Metacrinus rotundus* Carpenter (ligament collagenous fibers [ct] penetrate galleried stereom [gs] (G, H)). I–L. Labyrinthic stereom: a schematic drawing (I, adapted from Smith, 1980a); muscle attachment area on demipiramids of an echinoid *Prionocidaris baculosa* Lamarck (J, K); muscle attachment areas (m) linked to labyrinthic stereom with needle-like projections on a brachial of crinoid *Metacrinus rotundus* Carpenter (L).

Scale bars = 100 μm (A–D) and 50 μm (E–L), respectively.

(ii) Moderately fine and open meshwork (pores < ~16 μm, trabeculae < ~7 μm), common with thorns and microspines, is associated with muscle insertions (e.g., echinoid tubercles; ambulacral muscle pits of asteroids; brachial muscular facets in crinoids). It is also present (though without needle-like projections) in the autotomy planes of crinoid ossicles where nonpenetrative short collagenous fibers splay out;

(iii) Coarse meshwork (trabeculae > ~7 μm) is usually open textured and it is constructional. It commonly occurs on the internal and external surfaces of plates in all groups of extant echinoderms. It also forms the so-called uniform type of teeth in some ophiuroids. In deep-sea crinoids (hyocrinoids and holopodids) and echinoids (echinothuroids), this stereom may be associated with collagenous and muscle fibers (though its trabeculae are usually moderately fine or fine). It is also observed in muscle insertion areas of calcareous ring elements in some holothurians (though its trabeculae may be moderately fine).

3.4 Fascicular Stereom

Branching, more or less parallel and dense trabecular rods without continuous galleries. Rods are connected with each other by trabecular branching and small trabecular struts. Pores are irregular to elongate (Figure 5A, B). This stereom is commonly formed during growth predominantly developed in one direction (e.g., spines of some echinoids, asteroids and ophiuroids; ambulacral plates of asteroids; lateral surfaces of crinoid plates; buttressing in clypeasteroid tests; tubercle mamelons of echinoids; articulation pegs of ophiuroids; spires in body wall ossicles of holothurians).

3.5 Laminar Stereom

Thin sheet-like retiform layers separated by thin and more or less perpendicular pillars. The separation distance between sheets is commonly greater that the thickness of individual sheets (Figure 5C, D). It is used for the general plate construction where a thin layer of stereom is laid down over a broad area (e.g., inner plate margins and spines of some echinoids; ophiuroid tegmen/disc plates; body wall ossicles in some holothurians).

3.6 Retiform Stereom Layer

A layer of interconnecting trabeculae in one plane only. Its thickness is smaller than the maximum diameter of its pores, which are closely spaced

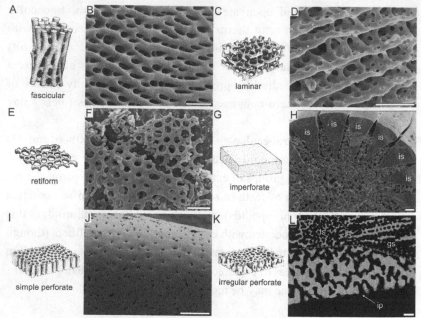

Figure 5 Different types of stereom fabrics in recent echinoderms.
A, B. Fascicular stereom: a schematic drawing (A, adapted from Smith, 1980a);
brachial latera of a crinoid *Hypalocrinus naresianus* Carpenter (B).
C, D. Laminar stereom: a schematic drawing (C, adapted from Smith, 1980);
spine (area near the milled ring) of *Prionocidaris baculosa* Lamarck (D).
E, F. Retiform stereom: a schematic drawing (E, adapted from Smith, 1980);
body wall ossicle of a holothurian *Psolus phantapus* Strussenfelt (F).
G, H. Imperforate stereom: a schematic drawing (G, adapted from Smith, 1980);
fractured echinoid spine of *Psammechinus miliaris* Müller showing septa
composed of imperforate stereom (is) (H). I, J. Simple perforate stereom:
a schematic drawing (I, adapted from Smith, 1980); columnal latera of
Hypalocrinus naresianus Carpenter (J). K, L. Irregular perforate stereom:
a schematic drawing (K, adapted from Smith, 1980); test plate of *Psammechinus
miliaris* Müller showing a contact between galleried (gs), labyrinthic (ls) and
irregular perforate stereom (ps) (L). Scale bars = 50 μm (A–L).

and may be linearly or irregularly arranged (Figure 5E, F). It commonly
forms muscle attachment platforms (e.g., areole of primary tubercles in
echinoids; muscle flanges on ambulacral plates in asteroids; brachial muscle
flanges in crinoids; vertebral muscle flanges in ophiuroids). It also forms
body-wall ossicles of holothurians.

3.7 Imperforate Stereom Layer

Solid layer without any perforations (Figure 5G, H). This stereom is observed on the external surfaces and articulation surfaces of plates in all extant echinoderm classes (e.g., septa of echinoid spines; larger tubercles of echinoids; tips of the so-called compound teeth in some ophiuroids; fulcral ridges). It is formed via secondary infilling of calcite on previously formed open stereom.

3.8 Perforate Stereom Layer

Dense crust, usually thicker than the maximum diameter of the pores that perforate it. Pores may be unbranched, and more or less perpendicular to the crust layer (the so-called simple perforate stereom; Figure 5I, J), irregularly arranged and branched (the so-called irregular perforate stereom; Figure 5K, L) or dense and running in three directions in a multilaminar layer (the so-called microperforate stereom). This stereom is commonly observed on the external surfaces of plates in all extant echinoderm classes (e.g., echinoid and crinoid plates; ophiuroid tegmen and arm plates; marginal plates of asteroids; articulation surfaces in different plates, including epiphyses and rotulae, and tubercle mamelon in echinoids; vertebral articulation pegs in ophiuroids; articulation ridges of brachial and columnal surfaces of crinoids; articulation pegs in asteroids).

3.9 Other Types of Biomineralized Structures in Echinoderms

Echinoderms may also produce other types of biominerals, which may differ significantly from the aforementioned types of stereom microarchitecture. For instance, an echinoid tooth is a highly complex structure that is composed of many individual elements (primary plates, lamellae, needles, secondary plates, prisms, and carinal processes; Figure 6A, B). These elements are solidified by a polycrystalline calcite matrix that is significantly enriched in magnesium (Schroeder et al., 1969). This matrix displays high hardness and elastic modulus, and, together with the specific arrangement and morphology of plates and needles, is thought to play a key role in grinding and self-sharpening functions (Ma et al., 2009).

The structure and appearance of the cidaroid cortex, an outer layer of mature spines, is also different. It is composed of two layers: the inner one consisting of microspines revealing typical conchoidal fracture, and the outer one forming spinules with lateral protuberances displaying unusual irregular polycrystalline structure (Figure 6C, D; Gorzelak et al., 2017a). Fully developed spines in cidaroids are not covered in epidermis, and thus are commonly encrusted with algae and epizoans (David et al., 2009), which probably provide them a camouflage and a first line of defence against predators.

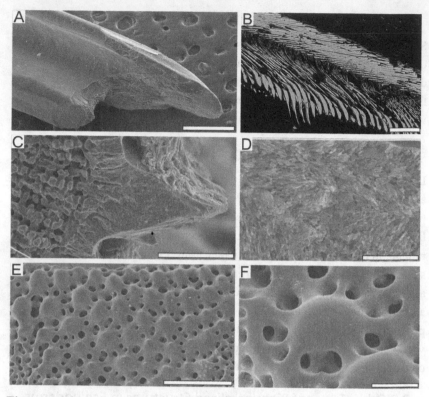

Figure 6 Other types of biominerals in recent echinoderms. A. Grinding tip of
an echinoid tooth of *Paracentrotus lividus* Lamarck. B. Cross section of an
echinoid tooth of *Paracentrotus lividus* Lamarck showing primary plates with
lamellae (pp + l), secondary plates with needles (sp +n) and carinal processes
(cp). C. Cortex of a spine of cidaroid *Prionocidaris baculosa* Lamarck with
enlargement (D). D. Enlargement of a fine irregular polycrystalline structure
from the outermost cortex. E. Enlarged peripheral trabeculae on the dorsal arm
plate of *Ophiocoma wendtii* Müller and Troschel with enlargement (F).
F. Enlargement of a single "lens." Scale bars = 0.5 mm (A), 100 μm (B), 200 μm
(C, E), 20 μm (D), 50 μm (F), respectively.

Other interesting biomineralized tissue observed in echinoderms are the so-
called enlarged peripheral trabeculae (EPTs) (Figure 6E, F). Sometimes they
form a regular array of lens-like extensions of stereom. Aizenberg et al. (2001)
demonstrated that these structures in ophiuroid *Ophiocoma wendtii* Müller and
Troschel are able to focus light onto presumably light-sensitive nerve bundles,
and thus are part of the complex photoreceptor system. These structures have
a unique morphology, resembling some flat-convex eyepiece Huygens-type
lenses in cross sections, minimizing spherical aberration, and characteristic

crystallography eliminating birefringence (orientation of the calcite crystallographic *c*-axis lies parallel to the lens axis; Gorzelak et al., 2017c). Furthermore, thanks to their unique nanostructural design, resembling some Guinier–Preston zones, they exhibit a high mechanical strength (Polishchuk et al., 2017). However, a recent study using immunohistochemical methods to recognize markers specific for some opsins, suggested that the functional significance of these structures may be different (Sumner-Rooney et al., 2018). More specifically, it has been shown that the photoreceptors associated with vision in *O. wendtii* are distributed in pores surrounding the EPTs. According to these latter authors, the convex morphology of the "microlenses" may just increase the acceleration of the dispersion of chromatophores, acting as filters covering the photoreceptors, producing a more dramatic color change in animals, thus leading to their more effective phototaxis. These authors, however, mentioned that they labeled the reactivity of two types of opsins. Given the observed high opsin diversity in echinoderms, it cannot be fully excluded that other photoreceptor types may be associated with "microlenses." Most recently, however, Sumner-Rooney et al. (2020) confirmed that *O. wendtii* displays a remarkable spatial vision thanks to their extraocular photoreceptors located in pores surrounding the EPTs, fields of view of which are restricted by chromatophores. Notably, Márquez-Borrás et al. (2018) argued that the EPTs are common in shallow-water species of the genus *Ophionereis* but are absent in forms living in caves (stygofauna). This suggests that these structures may be at least indirectly involved in photoreception.

Most surprisingly, apart from the calcite biominerals produced by echinoderms, certain species of molpadiid holothurians can also produce ferrous phosphatic and opal dermal granules (Lowenstam and Rossman, 1975). They are 10–280 μm in diameter. Their surface is usually smooth but in cross section they reveal a layered structure. Their function remains unknown.

4 Stereom of Fossil Echinoderms: Preservation Style and Observational Methods

Studies on stereom microstructure in fossil echinoderms are rather difficult. This is due to the fact that diagenesis commonly leads to a significant obliteration or loss of the primary stereom microarchitecture. Occasionally, calcite stereom may be replaced or thinly coated by silica, iron oxides, glauconite or phosphates (Figure 7A–C). Such ossicles can be extracted from rocks using ~10% acetic acid and/or HCl, and their three-dimensional stereom can be then observed directly under SEM (e.g., Pisera 1994; Clausen and Smith, 2008; Salamon et al., 2015). Echinoderms can be also preserved as molds (Figure 7D). In this case, fine sediments or pyritic framboids fill pores space in ossicles and then, if the original

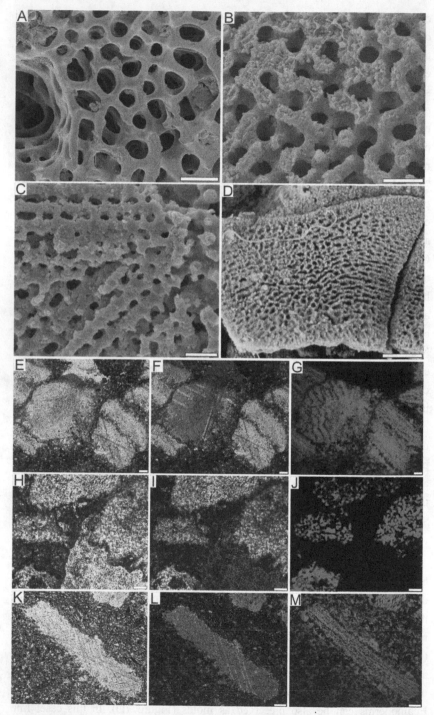

Figure 7 Different types of preservation of stereom microstructure.
A. Diagenetically unaltered stereom of an isocrinid brachial preserved as calcite

Caption for Figure 7 (cont.)

sealed with clays (Middle Jurassic, Łuków, Poland). B. Silicified stereom of a holocrinid columnal (Lower Triassic, Svalbard). C. Isolated phosphatized coronate echinoderm ossicle obtained after acid etching (Upper Ordovician, Mójcza, Poland). D. Latex cast of a cinctan *Graciacystis* Zamora, Rahman and Smith (middle Cambrian, Spain) preserving traces of stereom (photo courtesy of Samuel Zamora). E–M. Diagenetically altered stereom of the middle Cambrian echinoderms (from Morocco and Spain) preserved as recrystallized calcite (after Gorzelak and Zamora, 2013). E–G. Photomicrographs of integument plate of *Protocinctus* Rahman and Zamora under optical (E), polarizing (F) and cathodoluminescence (G) microscopy. H–J. Photomicrographs of an ambulacral plate of *Stromatocystites* Pompeckj under optical (H), polarizing (I), and cathodoluminescence (J) microscopy. K–M. Photomicrographs of an integument plate of *Protocinctus* Rahman and Zamora under optical (K), polarizing (L) and cathodoluminescence (M) microscopy. Scale bars = 20 µm (A–C), 500 µm (D), 100 µm (E–M), respectively.

calcite stereom is dissolved during diagenesis and/or weathering, a negative replica of the stereom is preserved, which can be observed thanks to a vacuum embedment of epoxy plastic into cavities and the digestion of rock matrix in HF acid (Zapasnik and Johnson, 1984). A bath in a solution of oxalic acid to produce a fully cleaned natural mold is also recommended (Zamora et al., 2013).

Most commonly, however, the porous ossicles tend to be filled by a secondary cement and/or some sediment which obscures the primary three-dimensional morphology of the stereom (Gorzelak et al., 2016). If the infill is composed of unconsolidated clay/mud, it can be simply removed by standard micropaleontological methods (paraffin or Quaternary 0). If the infill is strongly cemented and noncalcareous (e.g., siliceous), it can be dissolved by using HF acid, which additionally transforms the primary calcite skeletons into CaF_2 (Sevastopulo and Keegan, 1980). Unfortunately, echinoderm ossicles are usually filled by sparry calcite, and undergo transformation from unstable high-Mg calcite to a single monocrystal of stable low-Mg calcite. In this case, the stereom can be hardly recognizable (Figure 7E, F, H, I, K, L). Sometimes it can be detected after etching with formic dilute acid or under thin sections after staining with alizarin red (e.g., Dickson, 1966; Laphan et al., 1976; Smith, 1990). However, these methods are not effective if the chemical composition of the cement is similar to that of the stereom. Recently, a promising method, namely cathodoluminescence (CL), for investigating the stereom microstructure in fossil echinoderms under thin sections was introduced (Gorzelak and Zamora, 2013, 2016; Gorzelak et al., 2014b; Gorzelak et al., 2017b; Gorzelak, 2018). Cathodoluminescence is the emission of photons from a material that is subject to an electron bombardment. In carbonates, it is largely controlled by Mn^{2+}, which represents the most important activator element, whereas Fe^{2+} is the most efficient quencher (e.g., Richter et al., 2003). During diagenesis, Mn^{2+} commonly substitutes two different lattice positions within the Mg-calcite skeleton of echinoderms (i.e., Ca^{2+} position in Mg-free calcite "main-structure" and Mg^{2+} position in magnesite-like "sub-structure") leading to an intense orange-red luminescence (Figure 7G, J, M). By contrast, the carbonate cement filling porous ossicles is typically a ferroan calcite, which is nonluminescent. As a consequence of these distinct differences in concentrations of trace elements (which may be very slight, i.e., at the level of few ppm), the primary stereom microstructure that is not readily visible under transmitted light (Figure 7E, F, H, I, K, L) or SEM, can be greatly enhanced by CL (especially under highly sensitive hot-cathode: Figure 7G, J, M).

5 Stereom Microstructure: A Powerful Tool for Assessing the Paleobiology and Paleoecology of Fossil Echinoderms

Early studies on stereom microstructure of fossil echinoderms already proved that there is a great potential of studying stereom for understanding the paleobiology and paleoecology of extinct echinoderms. For instance, Smith (1978, 1980b), on the basis of microstructural observations of ambulacral pores in fossil echinoids, demonstrated that the Paleozoic echinoids lacked tube feet, and that the earliest echinoids that developed these functionally important projections appeared in the Middle Jurassic. Another interesting example was provided by Berg-Madsen (1986) who demonstrated that the columnals of Cambrian pelmatozoans are composed of galleried stereom, which indicates that penetrative collagenous tissues similar to those present in the stem of recent crinoids were already developed in early echinoderms. Consistently, Gorzelak and Zamora (2013) found a significant differentiation in stereom design in three Cambrian echinoderm taxa (*Protocinctus* Rahman and Zamora, *Stromatocystites* Pompeckj, Dibrachicystidae Zamora and Smith), and concluded that the same constraints that affect stereom in recent echinoderms (nature of investing soft tissue), likely characterized the earliest echinoderms. Notably, galleried stereom indicative of collagenous penetrative tissue has been documented in the stalks of a number of Paleozoic crinoids (*Pisocrinus* De Koninck, *Barycrinus* Wachsmuth, *Gilbertsocrinus* Phillips) (Ausich, 1977, 1983; Donovan and Franzen-Bengtson, 1988; Riddle et al., 1988).

Stereom was also investigated to infer the flexibility and musculature in fossil echinoderms. For example, Ausich (1977) and then Sevastopulo and Keegan

(1980) provided convincing evidence that muscles in the crinoid arms had already developed in some mid-Paleozoic taxa. Likewise, Gale (1987), based on similar arguments from stereom microstructure, argued that musculature in asteroids did not appear earlier than by the late Carboniferous. Interestingly, Smith (1990) noted that the early ophiuroids, despite having primitive arm structure composed of alternately arranged ambulacral ossicles, already developed intervertebral muscular connections. More specifically, the latter author showed that ambulacral ossicles of the Middle Ordovician ophiuroid *Taeniaster* Billings reveal distinct flanges composed of fine labyrinthic stereom indicative of muscles. This stereom is especially well developed dorso-laterally, rather than ventrally, suggesting that the arm movement was mainly confined to the lateral plane.

An iconic example of using stereom microstructure to infer functional anatomy of extinct echinoderms relates to the enigmatic group of fossil invertebrates – stylophorans (Stylophora). Based on microstructural studies of their skeletons revealing well preserved stereom microstructure, Carlson and Fisher (1981) and Clausen and Smith (2005) showed that these Cambrian animals, linked by some authors with primitive chordates (Calcichordata; Jefferies, 1999), belong to echinoderms, and that they possessed a muscular locomotory stem (or muscular "arm"; see Lefebvre et al., 2019, Clark et al., 2020).

Recently, development of new imaging techniques (including cathodoluminescence microscopy) allowed me to interpret microstructures of fossil, even diagenetically altered, echinoderms with precision unknown so far. In this way, it was possible to use the remains of enigmatic groups of extinct echinoderms for considerations about their functional micromorphology with precision comparable to the studies of recent echinoderms. In the following, I provide a review of selected case studies.

5.1 Pleurocystitids

Pleurocystidae Jaekel is a group of "cystoids" (Rhombifera, Blastozoa) with a strongly flattened theca, two short feeding appendages and a distally tapered stalk not terminated by any holdfast (Paul, 1984). These animals are generally considered to represent a mobile epibenthos (Sumrall, 2000). It has been suggested that the stalk of these echinoderms was flexible and played important role in locomotion either due to the presence of mutable collagenous tissues penetrating the stereom of columnals (Brower, 1999) or thanks to the muscles occurring in the central canal (Donovan, 1989). Microstructural investigations of the stalks of these echinoderms by Gorzelak and Zamora (2016) revealed that the synarthrial articular facets of their columnals are composed of dense labyrinthic stereom consisting of thin and irregular trabeculae with needle-like projections, characteristic for muscle fibers. These data suggest that the stalk of pleurocystitids indeed played a role in locomotion, but it was mostly possible due to the presence of muscles penetrating stereom on the articular facets of columnals (Figure 8; for more details, see also Gorzelak and Zamora, 2016).

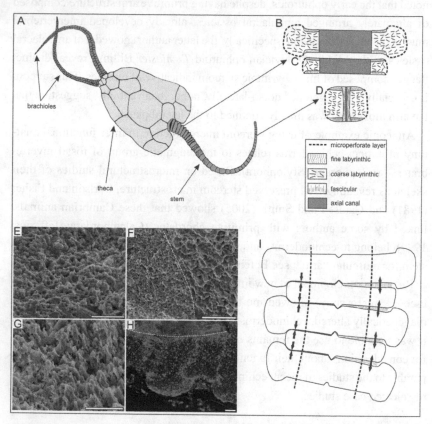

Figure 8 Functional morphology of Ordovician pleurocystitids from Canada (compiled and slightly modified from Gorzelak and Zamora, 2016). A. Pleurocystitid reconstruction. B–D. Stereom organization in outer proximal (B), inner proximal (C) and distal columnal (D). E. Labyrinthic stereom with needle-like projections on the articular face of a proximal columnal. F. Contact (dotted line) between coarse and fine labyrinthic stereom near the articular face of a proximal columnal. G. Fine labyrinthic stereom with needle-like projections on the articular face of a distal columnal. H. Longitudinally cross sectioned proximal columnals (the contact between cement is highlighted with dotted line) showing irregularly distributed pores. I. Functional model for pleurocystitid stem flexibility, muscles in relaxed (blue) and contracted (red) position, dotted lines delineate the position of the axial canal. Scale bars = 50 μm (E, G) 100 μm (F, H), respectively.

5.2 Ammonicrinids

The Devonian lecanocrinid crinoid *Ammonicrinus* Springer (Flexibilia), due to its peculiar morphology (including a spirally coiled stalk composed of moon-shaped columnals yielding spines), has long been the subject of interest for many paleontologists. In particular, the mode of life of these crinoids has become a subject of debate. According to the functional model by Bohatý (2011), ammonicrinids filtered food from a self-produced water flow mediated by the contraction of the stalk delivered by nonmuscular mutable collagenous tissues. This model, however, is not supported by actualistic data: in recent crinoids the mutable collagenous tissue contraction, allowing the bending of the entire stalk, proceeds very slowly (Ribeiro et al., 2011). Furthermore, living crinoids, despite having muscular articulations in the arms, are rheophilic suspension feeders and are not able to produce significant water flow, which would have been energetically costly. Microstructural investigations by Gorzelak et al. (2014b) documented a dense and fine superficial labyrinthic stereom on both sides of the lateral columnal enclosure extensions (LCEE) in ammonicrinids, suggesting that muscles might have also been present in their stalks. This is consistent with some other traditional proxies (including morphology of the facet and taphonomy). This finding reassessed the current views on the evolution of muscle connections in crinoids, and shed new light on the mode of life of ammonicrinids. In particular, these data show that muscles in some fossil crinoids might have not been necessarily confined to the arms as observed in recent crinoids. Furthermore, these results clearly suggest that the planispiral coil of the ammonicrinid stalk might have been possible not only due to the presence of mutable collagenous tissues, but also mostly due to the presence of muscles. The ability to quickly coil the stalk terminated with a small crown, was probably not connected with the ability to actively generate currents but, as in the case of spines, more likely constituted a protection against predators (Figure 9; for more details, see also Gorzelak et al., 2014b).

5.3 Uintacrinoids

Uintacrinoids (Uintacrinoidea Zittel) are enigmatic Cretaceous crinoids. These stalkless echinoderms possess a large globular theca without cirri and long arms (Hess and Messing, 2011). They display a wide geographic distribution and occur in a narrow stratigraphic range (late Santonian–early Campanian), thus they are considered biostratigraphically valuable. Given their unusual morphology, their paleoecology and mode of life have been the subject of different interpretations. These crinoids were classified as sessile or mobile epibenthos, semi-infauna, nekton or pseudoplankton. There were also different interpretations regarding

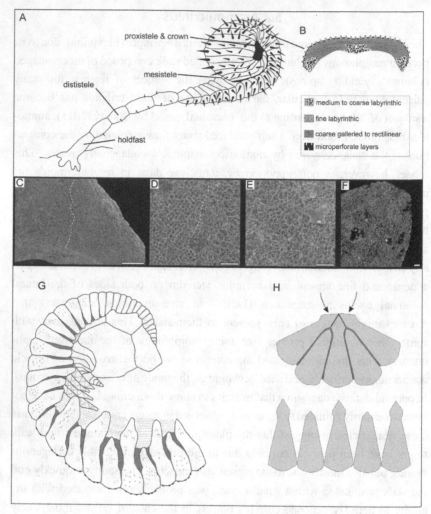

Figure 9 Functional morphology of Devonian ammonicrinids from Poland
(after Gorzelak et al., 2014b). A. Reconstruction of life position. B. Schematic
drawing of median columnal with distribution of stereom microfabrics.
C. Boundary (dotted line) between fine galleried (left) and fine labyrinthic
(right) stereoms. D. Higher magnification of (C). E. Higher magnification of
fine labyrithic stereom. F. Oblique-longitudinal section of the lateral columnal
enclosure extension showing medium to fine labyrinthic stereom.
G. Reconstruction of life position of *Ammonicrinus* Springer: enclosure of the
ammonicrinid crown in response to the external stimuli. H. Schematic sketches
of lateral columnal enclosure extensions of the mesistele in coiled (above) and
uncoiled (below) positions. Blue = inner ligaments, outer green = ligaments, red
= muscles. Scale bars = 500 μm (C), 100 μm (D, F), 50 μm (E), respectively.

their orientation toward the sea bottom: a mouth directed toward the bottom or the opposite (a review in Hess, 1999). Gorzelak et al. (2017b) critically evaluated previous arguments about the mode of life of these enigmatic crinoids, and provided new arguments from functional micromorphology, which support epibenthic lifestyle of uintacrinoids. In particular, microstructural investigations of two uintacrinoid genera (*Uintacrinus* Grinnell and *Marsupites* Mantell) revealed, in contrast to the previous studies, that their thecal and arm plates are massive and thick. They are composed of two stereom types: constructional galleried stereom consisting of thick trabeculae oriented in one direction, and fine galleried stereom, that is diagnostic for long collagenous ligaments binding adjacent plates. Throughout both stereom types, growth bandings composed of thick microperforate layers, which are known to increase plate strengthening in recent forms, are present. At the microscale level, the plate construction of uintacrinoids resembles those observed in recent benthic crinoids. However, it differs significantly from any pseudoplanktonic and pelagic forms (traumatocrinids and roveacrinids), which possess strongly perforated skeletons to increase buoyancy. Microstructural data, in connection with other arguments from taphonomy (such as preferential preservation of individuals arranged in a lateral position on the bedding surfaces in sediments of the so-called Konservat Lagerstätte) and functional "macromorphology" (presence of poorly developed muscular facets in the arms and inclusion of proximal parts of the arms in the theca, reducing their mobility), suggest that these crinoids were epibenthic. Their wide geographical range can be explained by the prolonged larval stage, which allowed their rapid dispersal over long distances (Figure 10; for more details, see also Gorzelak et al., 2017b).

5.4 Holocrinids

Holocrinids (*Holocrinus* Wachsmuth and Springer) are among the oldest Mesozoic crinoids. Owing to their phylogenetic significance, these crinoids have become the subject of much interest. Although the holocrinid stalk is similar to that of recent and fossil isocrinids, it lacks the flat synostosial articular facets at the distal side of their nodals, which are sites specialized for autotomy. They were thus long considered fully sessile. Preferential disarticulation of their stalks at the distal facets of nodals (Baumiller and Hagdorn, 1995) indicated, however, that *Holocrinus* was able to autotomize its stalk and likely relocate. It has been argued that the ability for stalk shedding in holocrinids evolved as an antipredatory adaptation against benthic predators during the so-called Mesozoic marine revolution (Baumiller et al., 2010). My microstructural investigations of holocrinid stalks (Gorzelak, 2018) revealed that they are mostly

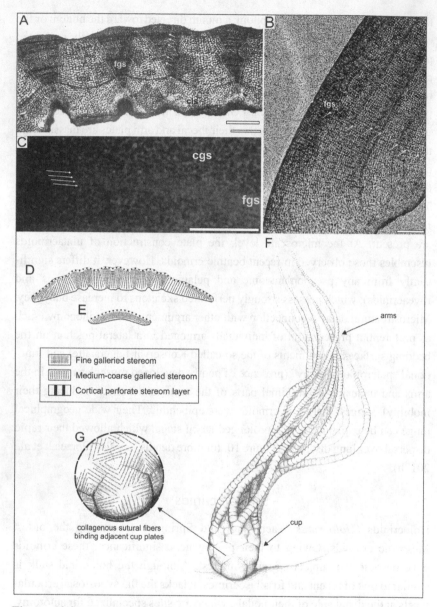

Figure 10 Stereom micromorphology and reconstruction of the life position of Late Cretaceous uintacrinoids (compiled and slighlty modified after Gorzelak et al., 2017b). A. Contact (delineated by dotted line) between fine (fgs) and coarse (cgs) galleried stereoms, locally transforming into coarse labyrinthic stereom (cls) in a thecal plate of *Marsupites testudinarius* Schlotheim. B. Thecal plate of *Uintacrinus socialis* Grinnel revealing fine (fgs) galleried stereom, and growth banding composed of perforate stereom layers (arrows). C. Thecal plate of *M. testudinarius* showing coarse galleried

Caption for Figure 10 (cont.)

stereom (cgs) separated by long galleries (highlighted by arrows) of fine regular stereom (fgs). D. Stereom organization in thecal plate of *Marsupites*. E. Stereom organization in a thecal plate of *Uintacrinus*. G. Reconstructions of the life position of *Marsupites* and soft tissue palaeoanatomy showing collagenous sutural fibers for binding adjacent plates (blue dotted lines). Scale bars = 0.5 mm (A, B), 0.25 mm (C), respectively.

composed of galleried stereom, diagnostic for long through-going ligaments binding adjacent columnals. On distal symplexial facets of nodals, however, fine and dense superficial synostosial stereom is observed. In recent crinoids, this stereom type is associated with short ligaments, these latter being characteristic for autotomy planes. Thus microstructural data support a previous hypothesis, based on taphonomic data only, suggesting ability for stalk autotomy in holocrinids (Figure 11; for more details, see also Gorzelak, 2018).

5.5 Scyphocrinoids

Scyphocrinoids (Scyphocrinitidae Jaekel) are a group of middle Paleozoic crinoids with highly unusual morphology. The stalk of these crinoids, up to about 3 m in length, is terminated distally with a globular, chambered structure, termed lobolith (Haude, 1972). For more than a century it has been commonly accepted that these crinoids were planktonic: they might have used a gas-filled lobolith as a buoyancy device which was moved along by surface currents, dragging along the filtration fan, which was suspended below. However, using evidence from skeletal micromorphology Gorzelak et al. (2020) demonstrated that the lobolith walls were not well adapted for preventing gas leaks and/or ingress of water (such as internal and external imperforate stereom layers). Instead, they are composed of homogenous constructional labyrinthic stereom. Furthermore, in plates from the distal side of the lobolith, wavy ridges and spines are observed. These data suggest that lobolith acted as a drag anchor rather than as a buoyancy device. In analogy to iceberg- and snowshoe-like strategies used by some brachiopods and mollusks, it can be hypothesized that the globular shape and the distally positioned microspines of loboliths might have served as adaptations for living on muddy bottoms. Notably, the planktonic mode of life of these crinoids was not supported by any biomechanical modeling which showed that the scyphocrinoid tow-net mode of feeding would have been highly ineffective (due to small relative velocities between the towed crown and the ambient water) (details in Gorzelak et al., 2020). It has been

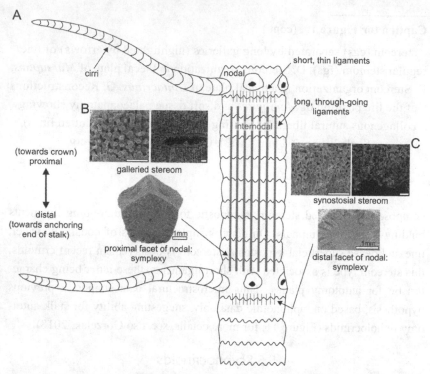

Figure 11 Inferred paleoanatomy of Triassic holocrinids (compiled and slightly modified after Gorzelak, 2018). A. Stalk composed of internodals connected to each other by long and thick through-going ligaments (marked in green) and nodals connected to the adjacent proximal facet of an internodal by thin and short ligaments (marked in blue). B. Galleried stereom of an internodal and a proximal facet of nodal with symplexial articulation. C. Irregular synostosial stereom of the distal facet of nodal with symplexial articulation and sealed lumen. Scale bars = 50 μm (B, C).

concluded that scyphocrinids maintained a feeding posture by extending the distal part of the stalk along the seafloor. Even in this recumbent posture, these crinoids could have occupied the highest epifaunal tier in the Paleozoic oceans (Figure 12; for more details, see also Gorzelak et al., 2020).

6 Conclusions

The echinoderm skeleton consists of numerous calcite plates with a unique mesh-like microstructure, termed stereom. A number of different stereom types have been identified in recent echinoderms. Despite evident relationships of stereom design to phylogeny and growth rate, the functional aspects of each stereom type

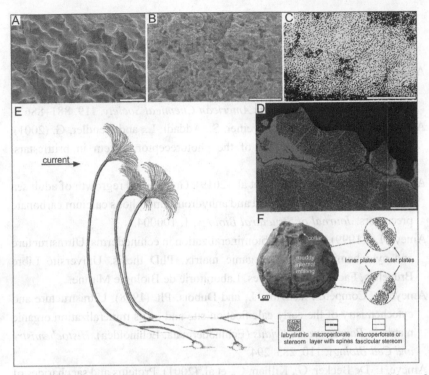

Figure 12 Stereom micromorphology and reconstruction of the life position of Late Silurian-Early Devonian scyphocrinoids (compiled and slightly modified after Gorzelak et al., 2020; reproduced with permission of the Licensor [Cambridge University Press] through PLSclear ref no 33074). A. Spiny ornamentation of a distal plate from the outermost lobolith wall. B. Relicts of the labyrinthic stereom inside a lobolith plate. C. External plates composed of coarse and dense labyrinthic stereom. D. Two layers of a plate lobolith composed of relicts of homogenous labyrinthic stereom. E-F. Reconstruction of scyphocrinoids as benthic crinoids (E) and distribution of stereom microstructure in scyphocrinoid loboliths (F). A, B = SEM images, C = optical microscope image, D = cathodoluminescence microscope image. Scale bars = 0.5 mm (A, D), 0.1 mm (B), 0.2 mm (C), respectively.

are the most important. For instance, galleried stereom, composed of a regular meshwork of deeply penetrating galleries, is commonly associated with penetrative collagenous tissues. In turn, an irregularly arranged, fine and dense, labyrinthic stereom (sometimes with needle-like projections) is very often linked to muscles. Given the close association between stereom microstructure and type of investing soft tissue, there is a great potential in determining paleoanatomy and paleoecology of extinct, enigmatic groups of echinoderms.

References

Aizenberg, J., Hanson, J., Koetzle, T. F., Weiner, S., and Addadi, L. (1997). Control of macromolecule distribution within synthetic and biogenic single calcite crystals. *Journal of the American Chemical Society*, **119**, 881–886.

Aizenberg, J., Tkachenko, A., Weiner, S., Addadi, L., and Hendler, G. (2001). Calcitic microlenses as part of the photoreceptor system in brittlestars. *Nature*, **412**, 819–822.

Alberic, M., Stifler C., Zou, Z., et al. (2019). Growth and regrowth of adult sea urchin spines involve hydrated and anhydrous amorphous calcium carbonate precursors. *Journal of Structural Biology*, **1**, 100004.

Ameye, L., (1999). Control of biomineralization in echinoderms: Ultrastructure and cytochemistry of the organic matrix. PhD thesis, Université Libre Bruxelles, Faculté des Sciences, Laboratorie de Biologie Marine.

Ameye, L, Compère Ph., Dille, J., and Dubois, Ph. (1998). Ultrastructure and cytochemistry of the early calcification site and of its mineralization organic matrix in *Paracentrotus lividus* (Echinodermata: Echinoidea). *Histochemistry and Cell Biology*, 110, 285–294.

Ameye, L. De Becker, G., Killian, C., et al. (2001). Proteins and saccharides of the sea urchin organic matrix of mineralization: Characterization and localization in the spine skeleton. *Journal of Structural Biology*, **134**, 56–66.

Ausich, W. I. (1977). The functional morphology and evolution of *Pisocrinus* (Crinoidea: Silurian). *Journal of Paleontology*, 51, 672–686.

Ausich, W. I. (1983). Functional morphology and feeding dynamics of the Early Mississippian crinoid *Barycrinus asteriscus*. *Journal of Paleontology*, **57**, 31–41.

Baumiller, T. K., and Hagdorn, H. (1995). Taphonomy as a guide to functional morphology of *Holocrinus*, the first post-Paleozoic crinoid. *Lethaia*, **28**, 221–228.

Baumiller, T. K., Salamon, M. A., Gorzelak, P., Mooi, R., Messing, Ch. G., and Gahn, F. J. (2010). Post-Paleozoic crinoid radiation in response to benthic predation preceded the Mesozoic marine revolution. *Proceedings of the National Academy of Sciences of USA*, **107**, 5893–5896.

Beniash, E., Addadi, L., and Weiner, S. (1999). Cellular control over spicule formation in sea urchin embryos: a structural approach. *Journal of Structural Biology*, **125**, 50–62.

Berg-Madsen, V. (1986). Middle Cambrian cystoid (*sensu lato*) stem columnals from 720 Bornholm, Denmark. *Lethaia*, **19**, 67–80.

Berman, A., Addaddi, A., and Weiner, S. (1988). Interactions of sea-urchin skeleton macromolecules with growing calcite crystals – a study of intracrystalline proteins. *Nature*, **331**, 546–548.

Berman, A., Hanson, J., Leiserowitz, L., Koetzle, T. F., Weiner, S., and Addadi, L. (1993). Biological control of crystal texture: a widespread strategy for adapting crystal properties to function. *Science*, **259**, 776–779.

Bohatý, J. (2011). Revision of the flexible crinoid genus *Ammonicrinus* and a new hypothesis on its life mode. *Acta Palaeontologica Polonica*, **56**, 615–639.

Bottjer, D. J., Davidson, E. H., Peterson, K. J., and Cameron, R. A. (2006). Paleogenomics of echinoderms. *Science*, **314**, 956–960.

Brower, J. C. (1999). A new pleurocystitid rhombiferan echinoderm from the middle Ordovician Galena group of northern Iowa and southern Minnesota. *Journal of Paleontology*, **73**, 129–153.

Carlson, S. J., and Fisher, D. C. (1981). Microstructural and morphologic analysis of a carpoid aulacophore. *Geological Society of America, Abstracts with Programs*, **13(7)**, 422.

Clark, E. G., Hutchinson, J. R., Bishop, P. J., and Briggs, D. E. G. (2020). Arm waving in stylophoran echinoderms: Three-dimensional mobility analysis illuminates cornute locomotion. *Royal Society Open Science* 7, 200191. http://dx.doi.org/10.1098/rsos.200191

Clausen, S., and Smith, A. B. (2005). Palaeoanatomy and biological affinities of a Cambrian deuterostome (Stylophora). *Nature*, **438**, 351–354.

Clausen, S., and Smith, A. B. (2008). Stem structure and evolution in the earliest pelmatozoan echinoderms. *Journal of Paleontology*, **82**, 737–748.

Cuif, J. P., Dauphin, Y., and Sorauf, J. E. (2011). *Biominerals and fossils through time*. Cambridge: Cambridge University Press.

David, B., Stock, S., De Carlo, F., Hétérier, V., and De Ridder, C. (2009). Microstructures of Antarctic cidaroid spines: Diversity of shapes and ecto-symbiont attachments. *Marine Biology*, **156**, 1559–1572.

Dickson, J. A. D. (1966). Carbonate identification and genesis as revealed by staining. *Journal of Sedimentary Research*, **36(4)**, 491–505.

Donnay, G., and Pawson, D. L. (1969). X-ray diffraction studies of echinoderm plates. *Science*, **166**, 1147–1150.

Donovan, S. K. (1989). The improbability of a muscular crinoid column. *Lethaia*, **22**, 307–315.

Donovan, S. K., and Franzen-Bengtson, C. (1988). Myelodactylid crinoid columnals from the Lower Visby Beds (Llandoverian) of Gotland. *GFF*, **110**, 69–79.

Dubois, Ph. (1991). Morphological evidence of coherent organic material within the stereom of postmetamorphic echinoderms. In S. Suga and H. Nakahara, eds., *Mechanisms and Phylogeny of Mineralization in Biological Systems*. Tokyo: Springer-Verlag, pp.41–45.

Dubois, Ph., and Chen, C. P. (1989). Calcification in Echinoderms. *Echinoderm Studies*, **3**, 109–178.

Dubois, Ph., and Jangoux, M. (1990). Stereom morphogenesis and differentiation during regeneration of fractured adambulacral spines of *Asterias rubens* (Echinodermata, Asteroidea). *Zoomorphology*, **109**, 263–272.

Gale, A. S. (1987). Phylogeny and classification of the Asteroidea (Echinodermata). *Zoological Journal of the Linnean Society*, **89**, 107–132.

Gorzelak, P. (2018). Microstructural evidence for stalk autotomy in *Holocrinus*: The oldest stem-group isocrinid. *Palaeogeography, Palaeoclimatology, Palaeoecology*, **506**, 202–207.

Gorzelak, P., Dery, A., Dubois, Ph., and Stolarski, J. (2017a). Sea urchin growth dynamics at microstructural length scale revealed by Mn-labeling and cathodoluminescence imaging. *Frontiers in Zoology*, **14**, 42, doi:10.1186/s12983-017-0227-8.

Gorzelak, P., Kołbuk, D., Salamon, M., Łukowiak, M., Ausich, W., and Baumiller, T. (2020). Bringing planktonic crinoids back to the bottom: Reassessment of the functional role of scyphocrinoid loboliths. Paleobiology, **46**(1), 104–122.

Gorzelak, P., Głuchowski, E., Brachaniec, T., Łukowiak, M., and Salamon, M. A. (2017b). Skeletal microstructure of uintacrinoid crinoids and inferences about their mode of life. *Palaeogeography, Palaeoclimatology, Palaeoecology*, **468**, 200–207.

Gorzelak, P., Głuchowski, E., and Salamon, M. A. (2014b). Reassessing the improbability of a muscular crinoid stem. *Scientific Reports*, **4**, 6049.

Gorzelak, P., Krzykawski, T., and Stolarski, J. (2016). Diagenesis of echinoderm skeletons: Constraints on paleoseawater Mg/Ca reconstructions. *Global and Planetary Change*, **144**, 142–157.

Gorzelak, P., Rahman, I.A., Zamora, S., Gąsiński, A., Trzciński, J., Brachaniec, T., and Salamon, M. A. (2017c). Towards a better understanding of the origins of microlens arrays in Mesozoic ophiuroids and asteroids. *Evolutionary Biology*, **44**, 339–346.

Gorzelak, P., Stolarski, J., Dery, A., Dubois, Ph., Escrig, S., and Meibom, A. (2014a). Ultra- and micro-scale growth dynamics of the cidaroid spine of *Phyllacanthus imperialis* revealed by ^{26}Mg labeling and NanoSIMS isotopic imaging. *Journal of Morphology*, **275**(7), 788–796.

Gorzelak, P., Stolarski, J., Dubois, P., Kopp, Ch., and Meibom, A. (2011). ^{26}Mg labeling of the sea urchin regenerating spine: Insights into echinoderm biomineralization process. *Journal of Structural Biology*, **176**, 119–126.

Gorzelak, P., and Zamora, S. (2013). Stereom microstructures of Cambrian echinoderms revealed by cathodoluminescence (CL). *Palaeontologia Electronica*, **16** (3), 32A.

Gorzelak, P., and Zamora, S. (2016). Understanding form and function of the stem in early flattened echinoderms (pleurocystitids) using a microstructural approach. *PeerJ*, **4**, e1820.

Grimmer, J. C., Holland, N. D., and Messing, C. G. (1984). Fine structure of the stalk of the bourgueticrinid sea lily *Democrinus conifer* (Echinodermata: Crinoidea). *Marine Biology*, **81**, 163–176.

Haude, R. (1972). Bau und Funktion der *Scyphocrinites*-Lobolithen. *Lethaia*, **5**, 95–125.

Heatfield, B. M. (1971). Growth of the calcareous skeleton during regeneration of spines of the sea urchin *Strongylocentrotus purpuratus* (Stimpson); a light and scanning electron microscope study. *Journal of Morphology*, **134**, 57–90.

Hess. H. (1999). *Uintacrinus* beds of the Upper Cretaceous Niobrara formation, Kansas, USA. In H. Hess, W. I. Ausich, C. E. Brett, M. J. Simms, eds., *Fossil Crinoids*. Cambridge: Cambridge University Press, pp. 225–232.

Hess, H., and Messing, Ch. (2011). *Treatise on Invertebrate Paleontology. Part T, revised, Echinodermata 2, volume 3, Crinoidea Articulata*. Lawrence, KS: Paleontological Institute, The University of Kansas.

Holland, N. D., Grimmer, J. C., and Wiegmann, K. (1991). The structure of the sea lily *Calamocrinus diomedeae*, with special reference to the articulations, skeletal microstructure, symbiotic bacteria, axial organs, and stalk tissues (Crinoidea, Millericrinida). *Zoomorphology*, **110**, 115–132.

Hyman, L. H. (1955). *The Invertebrates: Echinodermata IV*, New York: McGraw-Hill.

Jefferies, R. P. S. (1999). The calcichordate theory. *Science*, **236** (4807),1476.

Kołbuk, D., Dubois, Ph., Stolarski, J., and Gorzelak, P. (2019). Effects of seawater chemistry (Mg^{2+}/Ca^{2+} ratio) and diet on the skeletal Mg/Ca ratio in the common sea urchin Paracentrotus lividus. *Marine Environmental Research*, **145**, 22–26.

Kołbuk, D., Di Giglio, S., M'Zoudi, S., Dubois, P., Stolarski, J., Gorzelak, P. (2020). Effects of seawater Mg^{2+}/Ca^{2+} ratio and diet on the biomineralization and growth of sea urchins and the relevance of fossil echinoderms to paleo-environmental reconstructions. Geobiology, **18**, 710–724.

Kroh, K., and Smith, A. B., (2010). The phylogeny and classification of post-Palaeozoic echinoids. *Journal of Systematic Palaeontology*, **8**(2), 147–212.

Lapham, K. E., Ausich, W. I., and Lane, N. G. (1976). A technique for developing the stereom of fossil crinoid ossicles. *Journal of Paleontology*, **50**, 245–248.

Lefebvre, B., Guensburg, T. E., Martin, E. L. O., et al. (2019). Exceptionally preserved soft parts in fossils from the Lower Ordovician of Morocco clarify stylophoran affinities within basal deuterostomes. *Geobios*, **52**, 27–36.

Lowenstam, H. A., and Rossman, G. R., (1975). Amorphous, hydrous, ferric phosphatic dermal granules in Molpadia (Holothuroidea): Physical and chemical characterization, and ecologic implication of the bioinorganic fraction. *Chemical Geology*, **15**, 15–51.

Ma, Y., Aichmayer, B., Paris, O., Fratzl, P., et al. (2009). The grinding tip of the sea urchin tooth exhibits exquisite control over calcite crystal orientation and Mg distribution. *Proceedings of the National Academy of Sciences of USA*, **106**(15), 6048–6053.

Macurda D. B., and Meyer D. L. (1975). The microstructure of the crinoid endoskeleton. *The University of Kansas Paleontological Contributions*, **74**, 1–22.

Macurda D. B. (1976). Skeletal modifications related to food capture and feeding behavior of the basketstar. *Astrophyton. Paleobiology*, **2**, 1–7.

Märkel, K. (1986). Ultrastructural investigation of matrix-mediated biomineralization in echinoids (Echinodermata, Echinoidea). *Zoomorphology*, **106**, 232–243.

Márquez-Borrás, F., Solís-Marín, F. A., and Mejía-Ortíz, L. M. (2018). Troglomorphism in a brittlestar of the genus *Ophionereis* (Ophiuroidea: Ophionereididae) from Mexico. *Abstracts of the 16th International Echinoderm Conference, Nagoya, Japan*.

Martins, L., and Tavares, M. (2018). *Ypsilothuria bitentaculata bitentaculata* (Echinodermata: Holothuroidea) from the southwestern Atlantic, with comments on its morphology. *Zoologia*, **35**, e24573.

Medeiros-Bergen, D. E. (1996). On the stereom microstructure of ophiuroid teeth. *Ophelia*, 45, 211–222.

Moureaux, C., Pérez-Huerta, A., Compère, P., et al. (2010). Structure, composition and mechanical relations to function in sea urchin spine. *Journal of Structural Biology*, **170**, 41–49.

Nissen, H. U. (1969). Crystal orientation and plate structure in echinoid skeletal units. *Science*, **166**, 1150–1152.

Oaki, Y., and Imai, H. (2006). Nanoengineering in echinoderms: The emergence of morphology from nanobricks. *Small*, **2**, 66–70.

Okazaki, K. (1960). Skeleton formation of sea urchin larvae. II. Organic matrix of the spicule. *Embryologia*, **5**, 283–320.

Paul C. R. C. (1984). British Ordovician cystoids, part 2. *Monograph of the Paleontographical Society, London*, **136**, 65–153.

Pisera, A. (1994). Echinoderms from the Mójcza Limestone. *Palaeontologia Polonica*, **53**, 283–307.

Pearse, J. S., and Pearse, V. B. (1975). Growth zones in the echinoid skeleton. *American Zoologist*, **15**, 731–753.

Polishchuk, I., Brach, A. A., Bloch, L., et al. (2017). Coherently aligned nanoparticles within a biogenic single crystal: A biological prestressing strategy. *Science*, **358**(6368), 1294–1298.

Ribeiro, A. R., Barbaglio, A., Benedetto, C. D., et al. (2011). New insights into mutable collagenous tissue: Correlations between the microstructure and mechanical state of a sea-urchin ligament. *PLoS ONE*, **6** (9),e24822.

Richter, D. K., Goette, T., Goetze, J., Neuser, R. D., and Neuser, R. D. (2003). Progress in application of cathodoluminescence (CL) in sedimentary petrology. *Mineralogy and Petrology*, **79**, 127–166.

Riddle, S. W., Wulff, J. I., and Ausich, W. I. (1988). Biomechanics and stereomic microstructure of the *Gilbertsocrinus tuberosus* column. In R. D. Burke, P. V. Mladenov, P. Lambert and R. L. Parsley, eds., *Echinoderm Biology*. Rotterdam: A.A. Balkema, pp. 641–648.

Reich, M. (2015). Different pathways in early evolution of the holothurian calcareous ring? In S. Zamora, and I. Rábano, eds., *Progress in Echinoderm Palaeobiology*. Madrid: Inst. Geol. España: Cuadenos del Museo Geominero, **19**, pp. 137–145.

Roux, M. (1970). Introduction à l'étude des microstructures des tiges de crinoïdes. *Geobios*, **3**, 79–98.

Roux, M. (1971). Recherches sur la microstructure des pédonculés de crinoides post Paléozoiques. *Travaux du Laboratoire de paléontologie Orsay*, 1–83.

Roux, M. (1974). Les principaux modes d'articulation des ossicules du squelette des Crinoïdes pédonculés actuels. Observations microstructurales et conséquences pour l'interprétation des fossiles. *Compte Rendu de l'Académie des Sciences, Paris*, **278**, 2015–2018.

Roux, M. (1975). Microstructural analysis of the crinoid stem. *The University of Kansas Paleontological Contributions*, **75**, 1–7.

Roux, M. (1977). The stalk-joints of Recent Isocrinidae (Crinoidea). *Bulletin of the British Museum (Natural History). Zoology*, **32**, 45–64.

Roux, M., Messing, C. G., and Améziane, N. (2002). Artificial keys to the genera of living stalked crinoids (Echinodermata). *Bulletin of Marine Science*, **70**, 799–830.

Salamon, M. A., Gorzelak, P., Hanken, N. M., Riise, H. E., and Ferré, B. (2015). Crinoids from Svalbard in the aftermath of the end: Permian mass extinction. *Polish Polar Research*, **36**(3), 225–238.

Schroeder, J. H., Dwornik, E. J., and Papike, J. J. (1969). Primary protodolomite in echinoid skeletons. *Geological Society of America Bulletin*, **80**, 1613–1618.

Seto, J., Ma, Y., Davis, S. A., et al. (2012). Structure-property relationships of a biological mesocrystal in the adult sea urchin spine. *Proccedings of the National Academy of Sciences of USA*, **10**, 3699–3704.

Sevastopulo, G. D., and Keegan, J. B. (1980). A technique revealing the stereom microstructure of fossil crinoids. *Palaeontology*, **23**, 749–756.

Simms, M. J. (2011). Stereom microstructure of columnal latera: A character for assessing phylogenetic relationships in articulate crinoids. *Swiss Journal of Palaeontology*, **130**, 143–154.

Smith, A. B. (1978). A functional classification of the coronal pores of regular echinoids. *Palaeontology*, **21**, 81–84.

Smith, A. B. (1980a). Stereom microstructure of the echinoid test. *Special Paper in Palaeontogy*, **25**, 1–81.

Smith, A. B. (1980b). The structure, function, and evolution of tube feet and ambulacral pores in irregular echinoids. *Palaeontology*, **23**, 39–83.

Smith, A. B. (1982). The affinities of the Middle Cambrian Haplozoa (Echinodermata). *Alcheringa: An Australasian Journal of Palaeontology*, **6** (2), 93–99.

Smith, A. B. (1984). *Echinoid Palaeobiology*. London: George Allen and Unwin Ltd.

Smith, A. B. (1990). Biomineralization in echinoderms. In J. G. Carter, ed., *Skeletal Biomineralization: Patterns, Processes, and Evolutionary Trends*. New York: Van Nostrand Reinhold, pp. 413–443.

Sumner-Rooney, L. Rahman, I. A., Sigwart, J. D., and Ullrich-Lüter, E. (2018). Whole-body photoreceptor networks are independent of 'lenses' in brittle stars. *Proceedings of the Royal Society B: Biological Sciences*, **285** (1871), 20172590 doi: 10.1098/rspb.2017.2590

Sumner-Rooney, L., Kirwan, J. D., Lowe, E., and Ullrich- Lüter, E. (2020). Extraocular vision in a brittle star is mediated by chromatophore movement in response to ambient light. *Current Biology*, **30**, 319–327.

Sumrall C. D. (2000). The biological implications of an edrioasteroid attached to a pleurocystitid rhombiferan. *Journal of Paleontology*, **74**, 67–71

Weiner, S. (1985). Organic matrix-like macromolecules associated with the mineral phase of sea urchin skeletal plates and teeth. *Journal of Experimental Zoology*, **234**, 7–15.

Weiner, S., and Addadi, L. (2011). Crystallization pathways in biomineralization. *Annual Review of Materials Research*, **41**, 21–40.

Wilt, F. H. (1999). Matrix and mineral in the sea urchin larval skeleton. *Journal of Structural Biology*, **126**, 216–226.

Vidavsky, N., Addadi, S., Mahamid, J., et al. (2014). Initial stages of calcium uptake and mineral deposition in sea urchin embryos. *Proceedings of the National Academy of Sciences of USA*, **111**(1), 39–44.

Vidavsky, N., Addadi, S., Schertel, A., et al. (2016). Calcium transport into the cells of the sea urchin larva in relation to spicule formation. *Proceedings of the National Academy of Sciences of USA*, **113**(45), 12637–12642.

Yang, L., Killian, C. E., Kunz, M., Tamura, N., and Gilbert, P. U. P. A. (2011). Biomineral nanoparticles are space-filling. *Nanoscale*, **3**, 603–609.

Zamora, S., Rahman, I. A., and Smith, A. B. (2013). The ontogeny of cinctans (stem-group Echinodermata) as revealed by a new genus, *Graciacystis*, from the middle Cambrian of Spain. *Palaeontology*, **56**, 399–410.

Zapasnik, H. T., and Johnston, P. A. (1984). Replication in plastic of three-dimensional fossils preserved in indurated clastic sedimentary rocks. *Science*, **224**, 1425–1427.

Acknowledgments

This work was performed in the NanoFun laboratory co-financed by the European Regional Development Fund within the Innovation Economy Operational Programme POIG.02.02.00–00-025/09. I thank Dr Maciej Mazur (for providing Figure 1D), Dr Samuel Zamora (for providing Figure 7D), the Palaeontological Association (for permission to reproduce Figures 4A, E, I, 5 A, C, E, G, I, K), and the licensor – Cambridge University Press (for permission to reproduce Figure 12 through PLSclear ref no 33074). Bruno Ferré (Sotteville-lès-Rouen, France) and Mariusz Salamon (University of Silesia in Katowice, Poland) are acknowledged for useful remarks. Finally I would like to thank two anonymous reviewers for their comments, and an Editor-in-Chief Colin Sumrall for his invitation to contribute to this series. This paper was completed while I was a recipient of a grant from the National Science Centre (grant number 2016/23/B/ST10/00990).

Cambridge Elements ≡

Elements of Paleontology

Editor-in-Chief
Colin D. Sumrall
University of Tennessee

About the Series
The Elements of Paleontology series is a publishing collaboration between the
Paleontological Society and Cambridge University Press. The series covers the full
spectrum of topics in paleontology and paleobiology, and related topics in the Earth and
life sciences of interest to students and researchers of paleontology.

The Paleontological Society is an international nonprofit organization devoted
exclusively to the science of paleontology: invertebrate and vertebrate paleontology,
micropaleontology, and paleobotany. The Society's mission is to advance the study of the
fossil record through scientific research, education, and advocacy. Its vision is to be
a leading global advocate for understanding life's history and evolution. The Society has
several membership categories, including regular, amateur/avocational, student, and
retired. Members, representing some 40 countries, include professional paleontologists,
academicians, science editors, Earth science teachers, museum specialists, undergraduate
and graduate students, postdoctoral scholars, and amateur/avocational paleontologists.

Paleontological
S O C I E T Y

Elements of Paleontology

Elements in the Series

These Elements are contributions to the Paleontological Short Course on *Pedagogy and Technology in the Modern Paleontology Classroom* (organized by Phoebe Cohen, Rowan Lockwood and Lisa Boush), convened at the Geological Society of America Annual Meeting in November 2018 (Indianapolis, Indiana USA).

A full series listing is available at: www.cambridge.org/EPLY

Printed in the United States
By Bookmasters